Tick tock tick tock What's Up Croc?

Kim Michelle Toft

tick tock tick tock tick tock tick tock tick tock tick tock

Midnight strikes…

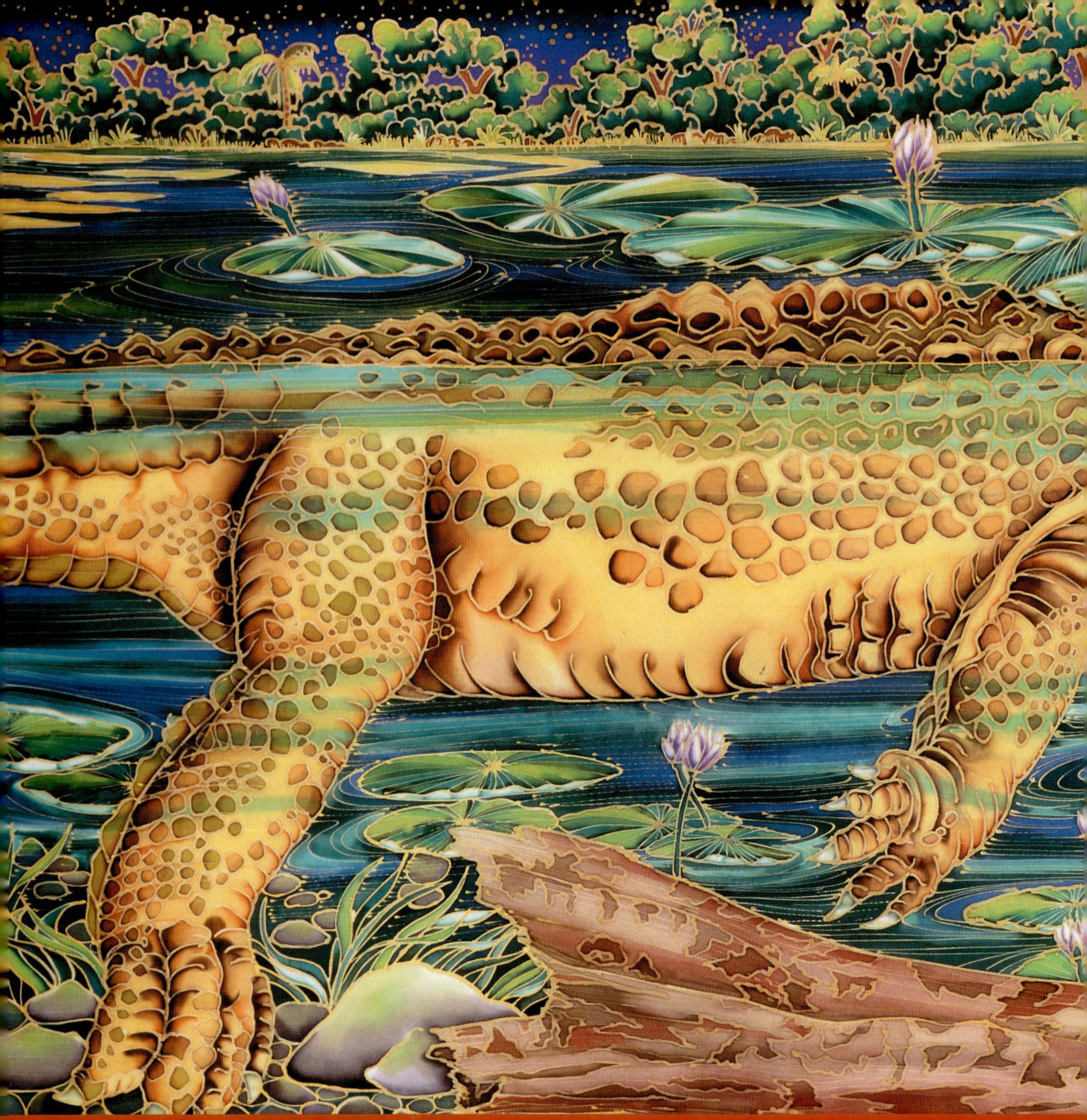

tick tock tick tock tick tock tick tock tick tock and a remarkable reptile,

tick tock tick tock tick tock tick tock tick tock from a time long gone by,

meanders ever so slowly *up* the lazy river in search of a tasty bite!

Barra or bird her favourite feast.

Belly full and floating free,

tick tock tick tock tick tock tick tock tick tock tick tock tick tock tick tock tick tock tick tock

she slithers sluggishly to the shore,
living statues basking in the midday sun.

Newborn cries sound the alert from a nest not far away.

tick tock tick tock tick tock tick tock tick tock tick tock tick tock tick tock tick tock tick tock tick

She scurries quickly to her mound,
buried treasure stirs inside.

Her babies hatch one by one and some need help from Mum!

Ferry rides to a weedy home, lovingly carried to the safety zone.

And when all are hidden and stashed away, she'll meander ever so slowly *down* the lazy river,

a remarkable reptile, survivor of time,
the last of the mighty dinosaurs.

tick tock tick tock tick tock tick tock tick tock tick tock tick tock tick tock tick tock tick tock tick

tick tock tick tock tick tock tick tock tick tock tick to   Midnight strikes...

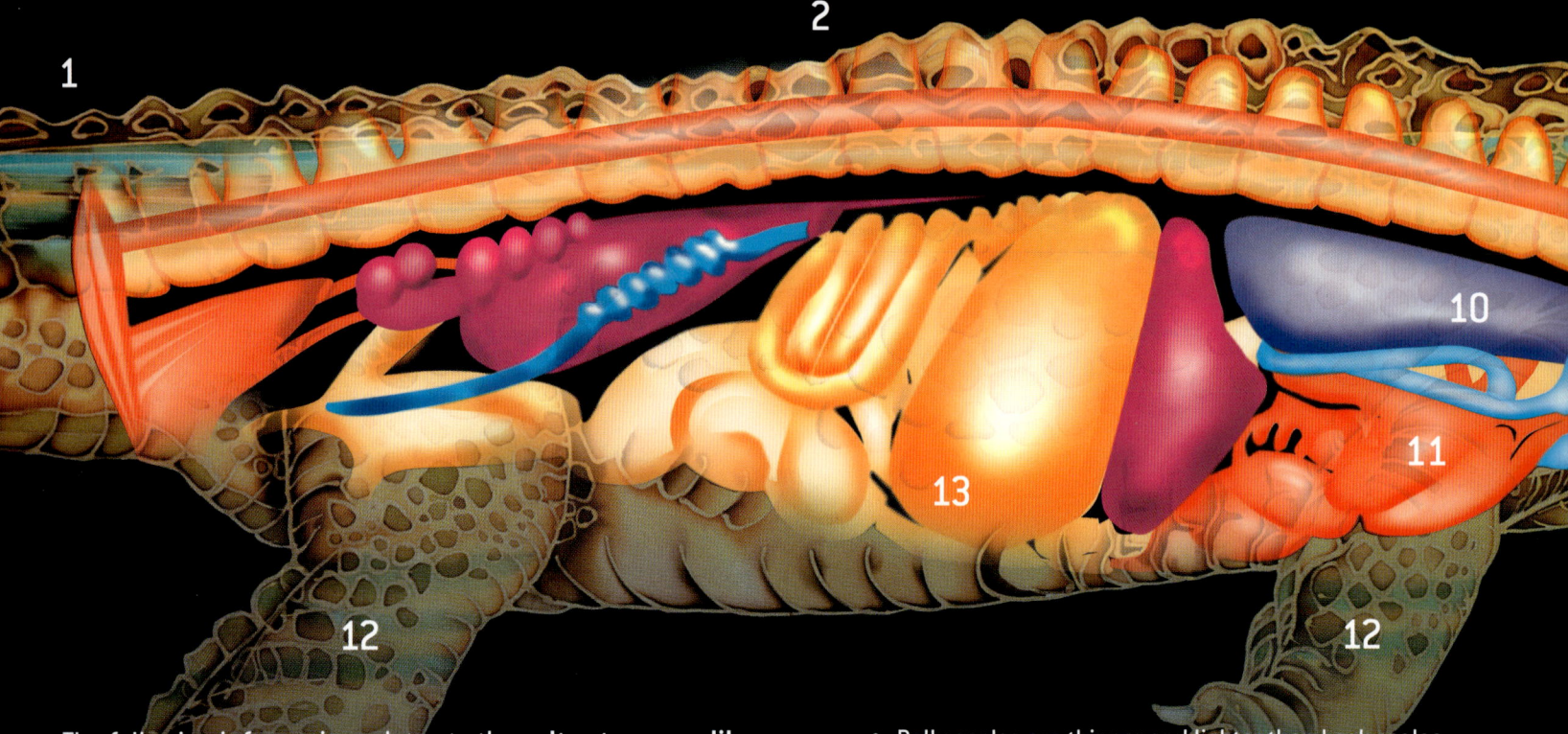

The following information relates to the **saltwater crocodile**, which will be referred to as the *croc* throughout the text.

The anatomy of a saltwater crocodile

1 Tail
- A muscular tail, flattened like a paddle, swishes from side to side to propel the *croc* through the water.
- Raised scutes along its tail are hard, not bony, and increase the surface area for swimming.
- Leaping straight out of the water, a *croc* seems to walk on its tail the same way a dolphin travels backwards on its strong tail.
- An adult *croc* can go without food for up to a year, living off fat stored in its tail and other parts of the body.

2 Body
- Weighs up to 1000 kilograms, approximately the weight of 3 small cars.
- Long, very heavy body, but very buoyant.
- Ability to submerge most of its body while having full use of its 3 senses — sight, hearing and smell.
- Has less protective armour on the neck and back so it is easier for it to bend its body when swimming.
- Small bones called *osteoderms* in the scutes along its back and tail protect the *croc* from bites by predators and other *crocs*.
- Entire body is covered with extremely tough, hard and horny material (or scales) set in a flat pavement fashion and used as armour.
- Belly scales are thinner and lighter than back scales.
- On land, a *croc* is cumbersome — sliding on its stomach, or walking with effort on its short legs.

3 Head
- When underwater all openings — inner and outer nostrils, ears, and throat valves — close up.
- Protected by bony plates that act like armour.
- A very tiny brain (more advanced than other reptiles) displays a complex repertoire of behaviours that enables the *croc* to learn things rather than act only on instinct.

4 Ears
- Only reptile with ear flaps
- Ears seal tight when underwater.
- Ears are high on the side of the head.
- No outer ears (as with humans) but special flaps of skin to protect the ears from water.

5 Eyes
- 3 eyelids
- A transparent third eyelid sweeps sideways across the eye for protection when underwater, and acts like swimming goggles to give clear vision
- Well-developed eyesight with a reflective area at the back of its eyes allows a *croc* to see at night.
- Eye pupils grow bigger the darker it gets.
- Eyes glow red in a spotlight at night making the *croc* an easy target for hunters.
- It is believed *crocs* may see in colour.

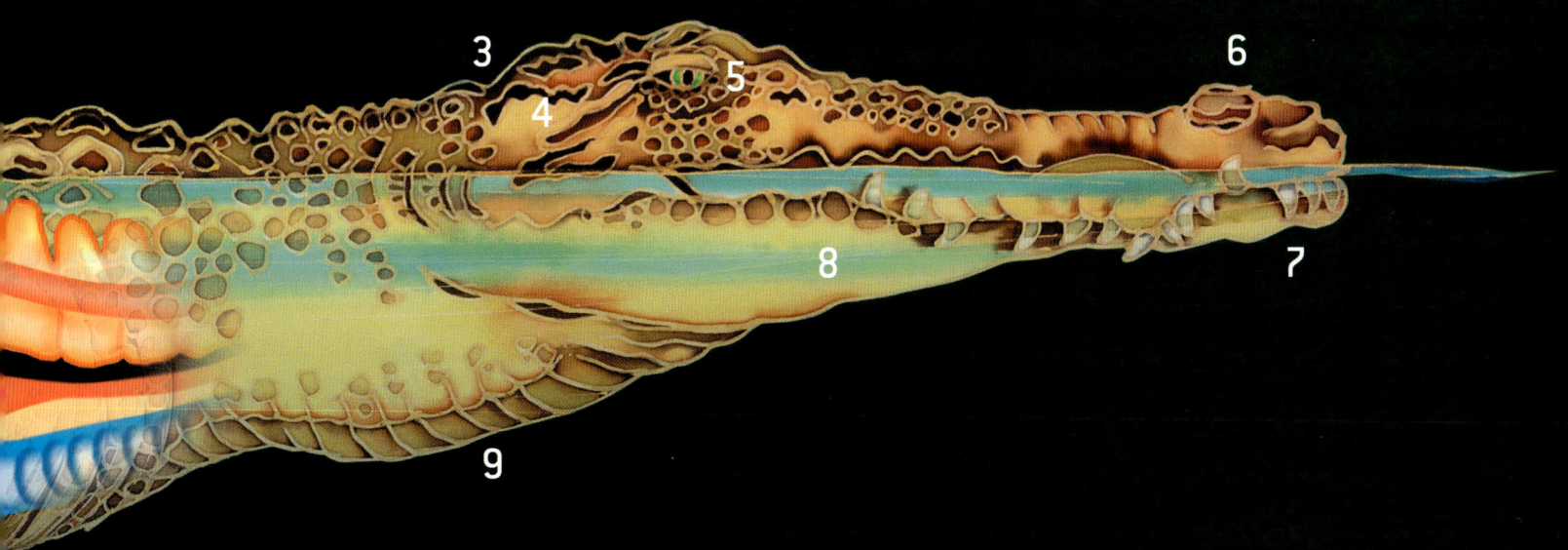

6   **Nostrils**
  - Internal nostrils seal watertight when submerged.
  - External nostrils, set high on the end of the snout above the water, allow a *croc* to breathe while almost totally submerged.

7   **Teeth**
  - 60 sharp pointed teeth
  - New teeth grow to replace broken teeth.
  - Adults (about 4 metres long) will have replaced each tooth about 45 times.
  - Several thousand teeth are grown during a lifetime.
  - Fourth tooth on the lower jaw sticks out when mouth closes.
  - Front jagged canines are used like knives to grip prey and blunt molar teeth at the back are used to crush prey.

8   **Mouth, tongue and jaw**
  - While basking in the sun, it opens its mouth to keep its brain from overheating.
  - The tongue is attached along the entire length of the floor of its mouth.
  - 40 large salt glands on the back of its tongue use fresh water to flush out (without using too much water) excess salt, especially from the kidneys.
  - 'Musk' glands under its lower jaw release sweet smelling oily perfume (green in colour) to attract the opposite sex.
  - Only the lower jaw moves downwards, while the top jaw doesn't move.
  - Sensitive scales on the sides of a *croc's* jaws help to feel and capture prey.

9   **Throat**
  - Two special flaps/valves at the back of the throat stop water entering the wind pipe (*trachea*).
  - When lunging at prey, these valves close to avoid the open mouth taking in water and drowning.

10 **Lungs**
  - Sac-like structures
  - Inhalation is achieved by moving the liver and other organs backwards.
  - In the water, the *croc* uses its lungs for buoyancy: floating higher when it breathes in; sinking lower when it breathes out.

11 **Heart**
  - Unlike other reptiles, a *croc's* heart has four chambers (as does a mammal) that allow the heart to pump more oxygen-rich blood to the brain during diving.
  - A *croc* can remain underwater for approximately one hour by slowing its heart to a single beat every 3 minutes.

12 **Legs**
  - Short legs positioned on the sides of its body, sprawl out when the *croc* is resting.
  - Legs are used to steer in the water.
  - Front legs have 5 toes with blunt claws but no webbing.
  - Back legs have 4 toes with webbing between 3 of them.
  - Hind feet provide power for a sudden rush through water or mud. They are also used to build its nest mound.

13 **Stomach**
  - Stomach has two parts: one for grinding food; the other for absorbing nutrients.
  - A *croc* swallows stones, which settle in the stomach to aid food digestion.

The environment and the croc

- Crocs can be found in oceans, fresh water and inland water courses such as rivers, wetlands and estuaries.
- Safety measures should always be adhered to when spending time in a *croc's* habitat.
  - Always take notice of signs and never swim where *croc*s may live.
  - Try to stand at least 5 metres away from the water's edge of *croc*-infected waters.
  - Avoid returning regularly to the same spot as *croc*s monitor behavioural patterns of prey.
- The Australian saltwater *croc* was once endangered. With hunting banned in 1974, their numbers have recovered and they are no longer considered 'endangered'. The decline in numbers in the past can be attributed to the destruction of habitat and the ruthless slaughter for their skins.
- *Croc*s are now farmed and harvested commercially for their leather and meat.
- Monitoring numbers is still important today as these resilient reptiles are our last link to prehistoric times.

A day in the life...

The following information relates to the **saltwater crocodile**, which will be referred to as the *croc* throughout the text.

Tick tock tick tock What's Up Croc?

- By lurking below the surface of the water, a *croc* becomes almost invisible. With only eyes and nostrils visible, a *croc* can move very close to its prey unnoticed.
- A *croc* can stay submerged for many hours with full use of its 3 senses — sight, hearing and smell.
- Watching every movement, a *croc* waits patiently for the perfect time to ambush and attack its prey.

Midnight strikes… and a remarkable reptile,

- Time stops for no living creature; the planet is always evolving.
- The same 52 weeks a year, 7 days a week, 24 hours a day, 60 minutes an hour and 60 seconds a minute propel all things into the future.
- Some creatures have defied the test of time, still roaming the planet today.
- The *croc* is the ultimate survivor, having lived on the planet for approximately 200 million years.

- The *croc* is the world's largest terrestrial carnivore (the largest land-based meat eater) growing to 7 metres in length and weighing a massive 1.5 tonnes.
- It is the world's largest reptile. A reptile is scaly-skinned and cold blooded with a bony skeleton and backbone.
- A *croc* is intelligent and well adapted to life in the water, swimming in both salt water and fresh water.
- A *croc* is also known as a 'salty'.

from a time long gone by, meanders ever so slowly *up* the lazy river in search of a tasty bite!

- The *crocodilia* group are related directly to the dinosaurs that ruled more than 200 million years ago.
- The first ancient crocodile, *Protosuchus*, meaning 'first crocodile', was small and probably no more than 1 metre long.
- During the Jurrassic period (200–145 million years ago) a colossal ten tonne crocodilian, *Sarosuchus*, or 'super crocodile', was one of the most ferocious predators, seen here battling a *Baryonyx*.
- Around 70 million years ago in the Cretaceous period, a gigantic *croc* called *Deinosuchus*, meaning 'terrible crocodile' arrived. *Deinosuchus* grew to 15 metres in length, similar in size to a *Tyrannosaurus rex* and twice as long as the largest crocodile today. *Deinosuchus* belonged to a group called the *Archosaurs* or 'ruling reptile'. Present day crocodiles are their only remaining relatives.

- A *croc* uses its muscular tail to propel it through the water.
- It swims along the coast, from river to river, and can be spotted hundreds of kilometres out to sea.
- Often found in estuaries, it is also known as an 'estuarine crocodile'.
- It can cruise the open waters for prey as it has no natural enemies. However, a young *croc* usually keeps to the shallows and waits for prey to pass.

Barra or bird her favourite feast.

- A *croc* will remain invisible under the water until it is ready to attack. It can either lunge at its prey dragging it down into the water or launch like a rocket to snatch its prey from the air.
- It is a fierce hunter that feeds on fish, crabs, snakes, birds and larger animals such as cattle, wallabies, pigs, dogs, and even humans!
- A *croc* does not chew its food, but swallows it whole. It grips its prey in its mouth and rolls its body to tear off big chunks. This is often referred to as the 'death roll'.
- The remains of the kill is often stored underwater in a cave or wedged under a log. This softens the flesh, making it easier to eat. The *croc* eats every part of its prey including the bones.
- During the wet season the *croc* is a more active hunter.
- A *croc* is a nocturnal hunter meaning it prefers to hunt at night.

Belly full and floating free,

- A *croc* swallows stones, which settle in its stomach. These stones help grind the food to aid digestion.
- With an extremely low metabolic rate, a large *croc* needs approximately 800 kilojoules per day, more if younger. (An average human requires around 6500 kilojoules per day.)
- A *croc* can go without food for long periods of time (2–12 months). During these lean times the *croc* lives off fat stored in its body.
- Although long and heavy, it has a very buoyant body, which allows it to float for many hours.
- A *croc* uses its lungs for buoyancy; floating higher when it breathes in and sinking lower when it breathes out.

she slithers sluggishly to the shore,
living statues basking in the midday sun.

- Like all reptiles, a *croc* depends on the sun to warm its blood. Only when a *croc's* blood is warm can it be active and hunt.
- After eating, the *croc* also warms its blood to help the stomach digest food properly.
- A *croc* is a social animal and often joins others in a group along the shore.
- A *croc* comes to land to:
  – bask in the sun to warm its blood
  – build and defend its nest mound
  – help babies reach the water after they have hatched
  – lunge at prey on the shoreline.

Newborn cries sound the alert,
from a nest not far away

- On average a mother *croc* will lay 50–100 eggs. Incubation of the eggs takes approximately 9–13 weeks.
- When a baby *croc* is ready to hatch, it will call out to its mother with high pitched grunts or cries from inside the egg.
- A mother *croc* guards her nest mound for up to 100 days. She will attack any predator that comes too close. Predators that prey on the eggs include goannas, wild pigs and birds.

She scurries quickly to her mound, buried treasure stirs inside.

- A female *croc* uses her clawed back feet to build her nest mound made of mud, reeds, grass and other vegetation.
- It is usually constructed at night, taking up to a week to build.
- As the vegetation rots, the heat produced helps the eggs to incubate.
- Nest mounds are built above the high tide level so the embryos inside the eggs do not drown.
- In the wild approximately 25% of eggs survive. Excessive flooding and predators such as large lizards, birds or mammals have an impact.
- When the mother *croc* hears her babies cry, she will race to her mound and scrap away nesting material to release them.
- The *croc* is not the best walker with a long heavy body and short legs positioned on the side. However, a smaller *croc* can gallop up to 17 kilometres per hour, but only for short distances.

Her babies hatch one by one and some need help from Mum!

- *Croc* eggs have hard shells and usually weigh between 110–140 grams (twice the weight of an average chicken egg).
- The temperature of the mound is important. At 30°C or lower the hatchlings will be mostly females. They develop slower and may take 95 days to hatch. At 33°C or above the hatchlings will be mostly males and develop much quicker at around 65 days. Temperatures in between will produce an even mix of both sexes.
- After 9–13 weeks, the baby *croc* hatches, no bigger than 30 centimetres long and weighing around 70 grams.
- Each hatchling has an eye tooth at the tip of its snout, which it uses to crack the egg shell open.
- If some eggs do not hatch, the mother *croc* will gently pick up the egg between her back teeth, shaking it for movement or sound. If there is none, she will eat the nutritious egg. If there is movement or sound, she will gently crack the egg allowing the baby *croc* to crawl along her tongue.

Ferry rides to a weedy home,
lovingly carried to the safety zone.

And when all are hidden and stashed away,
she'll meander ever so slowly
down the lazy river.

- A mother *croc* gently carries her hatchlings — in her mouth or on top of her head — to the water. Her ability to carry up to 15 hatchlings at a time is key to their survival.
- Hatchlings stay close together in the weedy shallows for about 2 months.
- A group of crocodilian young is called a 'pod'.
- In the 'nursery' the babies learn to swim and catch their own food. They start hunting straight away and feed on frogs, prawns, crabs and insects.
- A young *croc* faces many dangers from predators such as barramundi, long-necked turtles, freshwater crocodiles and sea eagles.

- A mother *croc* will carry her babies from the nest mound to a weedy area in the water. This helps protect the hatchlings from predators. She will make many trips to ensure all her babies are safely hidden in the water. This area is known as the 'crèche' or 'nursery'.
- A *croc* does not like to swim vigorously and is generally a lazy creature. It avoids strong waves wherever possible and prefers to drift with the tides in calmer water. Its long muscular tail propels it through the water, allowing it to use as little energy as possible
- As the *croc* swims underwater, it uses its transparent eyelids like swimming goggles to give clear vision.

a remarkable reptile, survivor of time, the last of the mighty dinosaurs.

- The *croc* is the world's most powerful creature.
- It is an energy-efficient, semi-aquatic predator that hunts with stealth.
- Throughout history, the *croc* was treated as a god, feeding on cakes and honey. It was also embalmed after death. Even a city called Crocodilopolis was built in Egypt in its honour.
- A *croc* is the closest living relative to the dinosaurs, defying time and still reigning supreme. It has changed little in the past 200 million years.
- A *croc* can live to 100 years of age and is more 'living legend' than 'living fossil'.

Midnight strikes…

- A female *croc* is ready to mate at 12 years of age, and a male *croc* at 16 years of age.
- To attract a mate a male *croc*:
   − lifts its head and bellows to warn off rival males
   − blows bubbles in the water to win the attention of female crocs
   − rubs its throat across a female's head and neck to release a sweet smelling oily perfume, green in colour. This is excreted from the 'musk' glands found under its lower jaw.
- A *croc* is an ardent lover, staying intertwined for up to 2 hours.
- A *croc* can mate several times during the day and may even mate with other partners, although the dominant male *croc* tries to prevent this. And so, the cycle of life continues.
- Throughout time, *crocs* have been worshipped, persecuted and protected. They are unrivalled as the most successful survivors in the history of the planet. *Crocs* remain our last living link to a prehistoric world.

This book is dedicated, as always, to my daughter Casey, who is now a young woman living in London.

First published 2010 by SILKIM BOOKS
PO Box 693 Ballina New South Wales 2478 Australia

© Kim Michelle Toft — text and illustrations

This book is copyright. Apart from any fair dealing for the purposes of private study, research, criticism or review, as permitted under the Copyright Act, no part may be reproduced without written permission. Enquiries should be made to the publisher.

www.kimtoft.com.au

Designed by Kim Michelle Toft
Graphic design and layout by Peter Evans
Printed by Everbest Printing Co. Ltd. China

Cataloguing in Publication Data
National Library of Australia

Author:	Toft, Kim Michelle
Title:	Tick tock tick tock what's up croc? / Kim Michelle Toft.
Edition:	1st ed.
ISBN:	978 0 9758390 6 5 (hbk.)
	978 0 9758390 7 2 (pbk.)
Target Audience:	For Children 9 years and under.
Subjects:	Crocodiles — Australia — Juvenile literature.
Dewey Number:	597.980994

Kim Michelle Toft books reflect her love of nature and the importance of its preservation. The unique illustrations are hand drawn with gold gutta directly onto white silk, then painted with brushes; using silk dyes.

Kim has spent her entire adult life on a beach somewhere in Australia. She currently lives on a picturesque beach in Northern New South Wales.